CONCEPTS AND TECHNIQUES IN MODERN GEOGRAPHY No. 4

SOME THEORETICAL AND APPLIED ASPECTS OF SPATIAL INTERACTION SHOPPING MODELS

by

Stan Openshaw
(University of Newcastle-upon-Tyne)

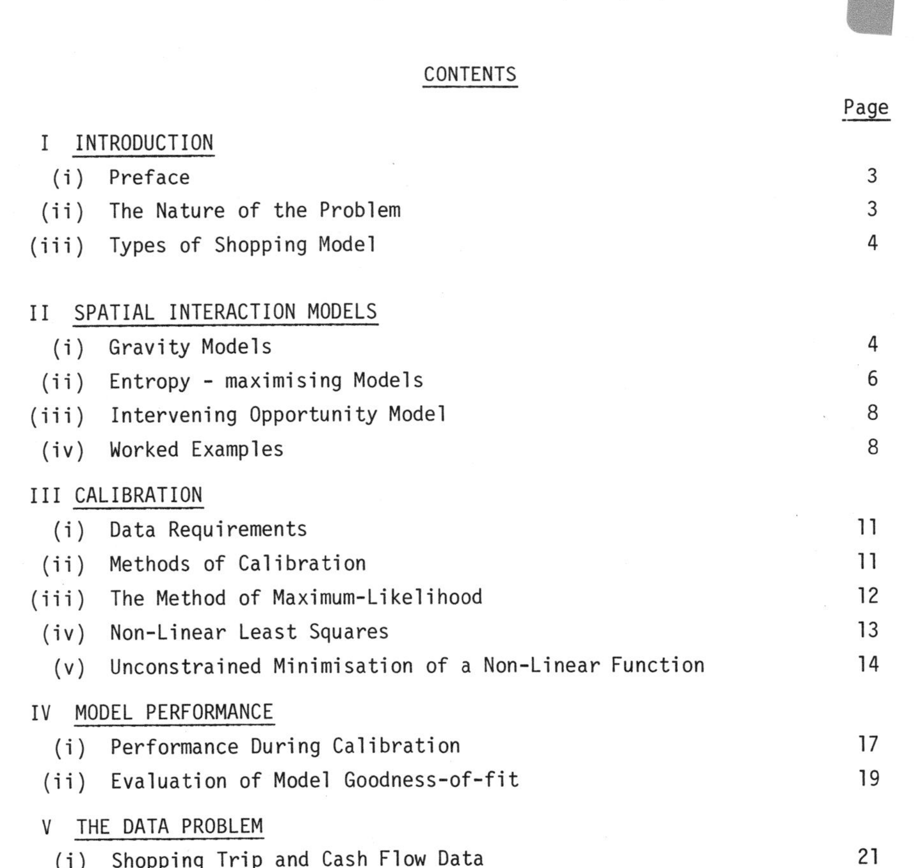

CONTENTS

Page

VI ASPECTS OF MODEL DESIGN

VII SHOPPING LOCATION-ALLOCATION OPTIMISING MODELS

VIII CONCLUSIONS

Acknowledgement

I would like to acknowledge the contributions made by C.J. Connolly as well as C.J. Collins and other members of Durham County Council Planning Department to this study.

THEORETICAL AND APPLIED ASPECTS OF SPATIAL INTERACTION SHOPPING MODELS

I INTRODUCTION

(i) Preface

The aim of this monograph is to describe and explain some of the characteristics of those shopping models which are members of the family of spatial interaction models. Particular attention is given to developing an understanding of these models in terms of selected aspects of model design, and of their application in practical situations concerning the description and forecasting of the spatial patterns of turnover. An introduction to the basic concepts of entropy-maximising and non-linear programming is also necessary to the development of a working knowledge of these models and an understanding of the problems associated with them. Finally, an attempt is made to show some of the directions in which shopping model building seems to be heading, and to illustrate the importance of developing a balanced theoretical and empirical approach.

(ii) The nature of the problem

Shopping models of some kind have often been used by geographers and planners to study the retail systems of an area. Such models can be considered geographical because of their emphasis on the spatial distribution of shopping trips as well as associated levels of sales, expenditures, and incomes. They also have an applied use in planning studies primarily as a means of forecasting the effects of changes in any of these distributions on the level of sales under the headings of impact, policy evaluation, and forecasting exercises.

The central problem in all retailing studies concerns describing the allocation of a person's present and future spending between various competing centres, or determining how much of a centre's turnover derives from people who live in a certain area.

An economist would approach this situation in terms of examining the micro behaviour of individuals, derive expressions for the utility of shopping in various centres, and consider the location of shops as a function of entrepreneurs maximising their profits in response to the demands for maximum utility from the consumers. Such an approach is aspatial and has obvious practical difficulties.

The traditional geographical approach is to change the scale of the problem replacing an emphasis on individuals by a set of zones formed by aggregating individual data to discrete spatial regions. A study area would usually be covered by a number of these zones. An attempt would now be made to find a relationship between the amount of expenditure available in each zone and the possible centres where the money could be spent. An alternative approach is to investigate the relationship between the level of turnover in a centre and factors which might be related to it, for example, levels of expenditure and accessibility. The various shopping models differ mainly in the way in which they handle these relationships. However, this approach is geographical only in the narrow sense of using data which has a spatial expression.

(iii) Types of Shopping Model

At least four major types of model may be identified (Harris, 1968). Some models relate the volume of retail trade directly to ambient purchasing power, for example, the EMPIRIC model expresses the rate of change in retail trade in terms of changes in the composition of the labour force and population levels (Hill, 1965). Likewise Lowry's 'model of metropolis' has a retail sector sub-model which distributes retail employment according to the strength of the market at each location (Lowry, 1964). These models tend to describe an excessively uniform distribution.

Other models are based on central place theory and involve developing regression equations to describe the numbers and sizes of retail establishments (Berry; 1965, 1967).

Recently the same multiple regression theme was used to describe the proportion of non-local expenditure going from one settlement to another as a linear function of relative attractiveness and relative accessibility (Lewis and Bridges, 1974).

Another class of models have been developed from an analogy with Newton's Law of Gravitation and more recently by a different analogy based on statistical mechanics. An alternative approach has been based on the intervening opportunity hypothesis. All these models may be grouped together under the blanket heading of Spatial Interaction Models since their prime concern is with movement patterns across space. As such they are of obvious geographical relevance, and the use of these models in a retail context forms the subject of the remaining sections of the monograph.

II SPATIAL INTERACTION MODELS

(i) Gravity Models

Gravity models are based on Newton's Law of Gravitation which states that the gravitational force, F_{ij}, between two masses, m_i and m_j, separated by distance c_{ij} is

$$F_{ij}(m_i, m_j, c_{ij}) = \frac{k \times m_i \times m_j}{c_{ij}^2}$$

where k is a constant.

The terms in brackets indicate the independent variables, although it has been common practice not to specify independent variables. However for the sake of clarity, it is essential to spell out those variables we are going to allow to vary. Thus if we have constant masses but variable c_{ij}, then the model would be written as $F_{ij}(c_{ij})$. If we wished to allow k to vary for given masses and distance variables as we would during calibration, then the model would be written as $F_{ij}(k)$. It is in terms of a calibration perspective that the independent variables for all subsequent models have been identified. It is left to the reader to carry out the changes necessary for a forecast situation when some of the variables may be held constant and others allowed to vary.

In a retailing situation the proprietor of an establishment would be interested in estimating the effectiveness of his trade in competition with similar traders in other centres near him. The gravity model hypothesis has a direct appeal and an early model using this analogy was developed by Reilly (1929) and modified by Converse (1930). This model assumes that shopping centres attract trade from intermediate places approximately in proportion to the sizes of the centres and in inverse proportion to the square of the distances to the intermediate place. Thus

$$\frac{Bij_1}{Bij_2} = \frac{Aj_1}{Aj_2} \times \left[\frac{Dij_2}{Dij_1}\right]^2 \qquad (1)$$

where Bij_1 and Bij_2 represent the trade attracted from the intermediate town i to towns j_1 and j_2 respectively (Styles, 1969).

This model of course restricts interaction to only two focal points. Nevertheless it is possible to define the market areas of two centres as a line of equal attraction between them. The breakpoint in distance units from centre j_2 is

$$\frac{Dj_1j_2}{1 + \sqrt{\frac{Aj_1}{Aj_2}}} \qquad (2)$$

where Dj_1j_2 is the distance between j_1 and j_2, and Aj_1 and Aj_2 are the populations of centres j_1 and j_2.

This catchment area approach was for a long time popular in planning (Pope, 1969), and a model like this was used in the Haydock Report (1964), although its only virtue seems to be that it provides a quick answer when computer facilities are not available.

No real progress was made until Huff (1962, 1964) proposed that

$$pij(\hat{n}) = \frac{Wj \times cij^{-\hat{n}}}{\sum_{j=1}^{m} Wj \times cij^{-\hat{n}}} \qquad (3)$$

$pij(\hat{n})$ is the probability that a resident in zone i would shop in j; Wj is a measure of shopping centre size; cij the travel time between i and j; $\hat{n}$ the independent variable which has to be estimated; the hat indicates that we will never know its true value only an estimated one with a known, or unknown, degree of variability; and m the number of zones.

This model was used to define market areas in terms of probability contours.

Lakshmanan and Hansen (1965) independently produced an equivalent model which by including an expenditure variable converted the probability pattern into cash flows:

$$Sij(\hat{a},Ai,\hat{b}) = ei \times Pi \times Ai(\hat{a},\hat{b}) \times Wj^{\hat{a}} \times cij^{-\hat{b}}$$
$$Ai(\hat{a},\hat{b}) = \left(\sum_{j=1}^{m} Wj^{\hat{a}} \times cij^{-\hat{b}}\right)^{-1} \qquad (4)$$

Sij() is the flow of retail expenditure from residential zone i to shopping centre j; ei is the mean expenditure per head in zone i; Pi is the population in zone i; Wj is the attractiveness of the shops in j; and $\hat{a}$ and $\hat{b}$ are the independent variables which have to be estimated.

However, these Newtonian gravity models have only a weak theoretical basis. Consider for example the justification for including the Ai() term in the previous model and a similar term in Huff's version. This parameter ensures that the constraint

$$\sum_{j=1}^{m} Sij(\hat{a},Ai,\hat{b}) = ei \times Pi \qquad (5)$$

is satisfied but it is not found in Newtonian physics.

This constraint ensures that the total sales generated by the residents of any zone is equal to their total retail expenditure; no more, no less.

(ii) Entropy-Maximising Models

It has been shown by Wilson (1967) that the gravity model can be given a sound theoretical explanation and that the Newtonian analogue can be replaced by a far more general technique for model building based on statistical mechanics. The Newtonian model's origin and destination mass variables are replaced by the entropy maximising method's emphasis on individual trips. The Newtonian derivation can be dropped, but the term 'gravity model' should be retained to describe those spatial interaction models which do not satisfy the requirements of the entropy-maximising approach. It is important to realise that these entropy-maximising models do not totally represent the whole field of spatial interaction modelling. They are in fact only one of several possible avenues of research although at present the dominant one.

It would be too great a digression to delve too deeply into the details of the entropy-maximising approach especially as an excellent introduction has been provided by Gould (1972). Yet this is such an important field that it is essential that the broader principles as they apply to shopping models are understood.

Suppose we have a system of residential zones (full of expenditure), shopping zones (full of shops), and a pattern of cash flows between these zones produced by a particular set of shopping trips. This same pattern of cash flows (a mesostate of the system) can be produced by a large number of different individual shopping trip combinations (micostates of the system). In fact any particular pattern of cash flows can be produced by

$$\frac{S!}{\prod_i \prod_j Sij!} \qquad (6)$$

different combinations of individual cash flows;

where $3! = 3 \times 2 \times 1$,

$$\prod_i \prod_j^{2} Sij = S_{11}! \times S_{12}! \times S_{21}! \times S_{22}!,$$

and S is the total of all shopping expenditures.
If each of these microstates is equally likely, then it seems reasonable to consider that the most likely pattern of cash flows, or mesostate, is the one which occurs most frequently. In fact the model which maximises equation (6) is

$$Sij = ei \times Pi \times Wj.$$

Not a very useful shopping model. It seems sensible, therefore, to try and develop a model which incorporates the contraint found in the Lakshmanan-Hansen model. This forms a macrostate constraint to which both mesostates and microstates should conform. To identify an entropy-maximising model based on this information we must maximise equation (6) subject to an equation (5) constraint for every origin zone, and produce

$$Sij = ei \times Pi \times Wj \times Ai$$

$$\text{where} \quad Ai = \left(\sum_{j=1}^{m} Wj\right)^{-1} \tag{7}$$

This is still not a very practical model as it incorporates no measure of locational choice. To include locational choice it turns out that we need an additional constraint which fixes the total amount spent on travelling to the shops, that is to say

$$\sum_{i=1}^{m} \sum_{j=1}^{m} Sij \times cij = \sum_{i=1}^{m} \sum_{j=1}^{m} Nij \times cij \tag{8}$$

where Nij is the observed cash flows between i and j.

We now maximise equation (6) subject to equations (5) and (8) to identify yet another entropy-maximising model, this time of the form

$$Sij(\hat{b},Ai) = ei \times Pi \times Wj \times Ai(\hat{b}) \times \exp(-\hat{b} \times cij)$$

$$\text{where} \quad Ai(\hat{b}) = \left(\sum_{j=1}^{m} Wj \times \exp(-\hat{b} \times cij)\right)^{-1}, \tag{9}$$

which is very similar to the gravity models previously identified but with a negative exponential cost function as the preferred form of distance deterrence function (Taylor, 1975). In a similar fashion it is possible to introduce additional information in the form of other macrostate constraints. Thus we may want the attraction term to reflect some notion of economies of scale, say $Wj^{\hat{a}}$, in which case the appropriate macrostate constraint would be

$$\sum_{i=1}^{m} \sum_{j=1}^{m} Sij \times \ln Wj = \sum_{i=1}^{m} \sum_{j=1}^{m} Nij \times \ln Wj \tag{10}$$

where ln is the symbol for natural logs.

It is important to realise that entropy-maximising is a model building technique and is not a hypothesis. Furthermore the form of the models that are produced is consistent with all the information we have declared through the macrostate constraints, but the model is also totally uncommitted in relation to what is unknown (Baxter, 1972). Indeed as the examples have shown, the definition of the macrostate constraints has an important effect on the form of the model since it is through these constraints that we incorporate theory into the models we are building. It is also a feature of this approach that the terms which occur in the equations have a physical significance in terms of their associate macrostate constraints. Thus in equation (9) the parameter $\hat{b}$ reflects the travel cost constraint of equation (8). Another comment relates to the fact that the 'individual' used to define these models is not the individual household or person but relates to an aggregate of these units over a spatial zone. Finally, the function we have been maximising, equation (6), is closely related to information theory and particularly the concept of entropy (a measure of uncertainty) and hence the derivation of the term entropy-maximising.

(iii) Intervening Opportunity Model

An important alternative to the gravity hypothesis is provided by the intervening opportunity model of spatial interaction. Stouffer (1940) suggested that the number of trips from an origin to a destination zone is proportional to the number of opportunities at the destination zone and is inversely proportional to the number of intervening opportunities. Wilson (1974, p 398) re-expresses this model so that there is a constant probability that a traveller will be satisfied at the next opportunity. Replacing trips by cash flows we get:

$$Sij[\mu,i]\ (\hat{L}) = ei \times Pi \times (\exp(-\hat{L} \times Aj[\mu-1,i]) - \exp(-\hat{L} \times Aj[\mu,i])) \qquad (11)$$

where $j[\mu,i]$ is the μth ranked destination zone j from i; and $Aj[\mu,i]$ is the number of opportunities up to and including $j[\mu,i]$ measured in terms of sales.

Thus the probability of a shopping trip from zone i going beyond any centre j (which is the μth ranked centre to i) is inversely proportional to the total number of opportunities nearer to i than this particular j, as well as the opportunities at j itself.

Wilson (1970) provides an entropy-maximising derivation of this model also but the argument is not so convincing. Nevertheless this model could be extended in many ways, for example, by the addition of an origin end constraint. In practice it has proven difficult to provide estimates of $\hat{L}$ because of numerical instability introduced by the exponential terms. Furthermore, the assumption that the opportunity acceptance probability function is globally determined and fixed for all zones does not make much geographical sense.

(iv) Worked Examples

A detailed discussion of how to build a shopping model is given by Wilson (1974, p 33-46) and this is not repeated here. Instead attention is focused on understanding what the calculation of these models involves. For this purpose let us assume a simple five zone study area and the data given in Figure 1. Reilly's model could be used to define the market area around zone 5 using equation (2). This involves estimating the breakpoint distances between zones 5 and 4, 5 and 3, and 5 and 1;

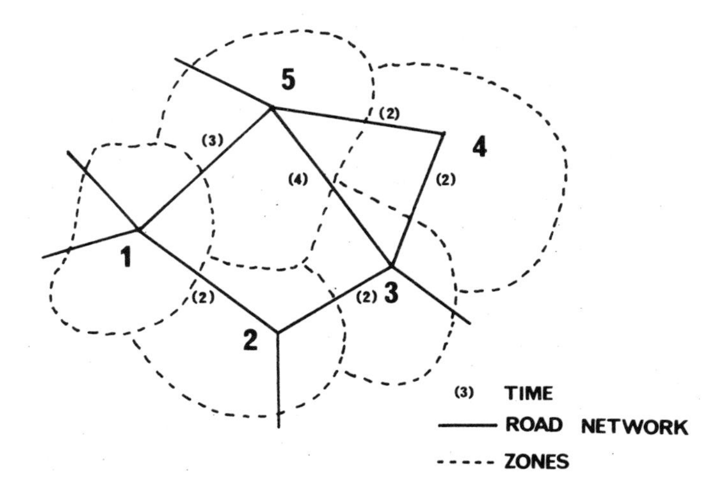

TO ZONE

		1	2	3	4	5
FROM ZONE	1	0	2	4	5	3
	2		0	2	4	6
	3			0	2	4
	4				0	2
	5					0

Cij

ZONE	Pi	ei	Wj
1	10	10	20
2	20	10	20
3	10	10	5
4	30	10	20
5	5	10	10

Fig. 1 Hypothetical data for worked examples

for 5 and 4 this breakpoint distance is $\dfrac{2}{1 + \sqrt{\frac{20}{10}}}$

or .83 units of time;

for 5 and 3 $\dfrac{4}{1 + \sqrt{\frac{5}{10}}}$ = 2.34 units of time;

and for 5 and 1 $\dfrac{3}{1 + \sqrt{\frac{20}{10}}}$ = 1.24 units of time.

Plotting these points on the network in Figure 1 would define the market area for zone 5.
The entropy-maximising model in equation (9) would calculate the cash flows between zones 1(i) and 3(j) as

$$S_{13} = e_1 \times P_1 \times W_3 \times A_1() \times \exp(-\hat{b} \times 4)$$

where $A_1() = (W_1 \times \exp(-\hat{b} \times 0) + W_2 \times \exp(-\hat{b} \times 2) + W_3 \times \exp(-\hat{b} \times 4) + W_4 \times \exp(-\hat{b} \times 5) + W_5 \times \exp(-\hat{b} \times 3))^{-1}$

If $\hat{b}$ = .1 then substituting values in this equation from Figure 1 would produce S_{13} = 28.27. Likewise, if $\hat{b}$ = .2 then S_{13} = 23.16, if $\hat{b}$ = .3 then S_{13} = 18.36, and, if $\hat{b}$ = .4 then S_{13} = 14.13

The intervening opportunity model in equation (11) is not quite as straightforward because of the additional subscripts. Let D_j be the number of opportunities at zone j which in terms of distance is μth ranked away from origin zone i, thus we get $D_j[\mu,i]$. The first step then is to rank the destination zones from an origin zone, say origin zone 1 (i=1), thus

Rank, μ, =	1st	2nd	3rd	4th	5th
zone j =	1	2	5	3	4

Now arrange the data in this order so that the number of opportunities is in μ order

$D_j[\mu,i]$	20	20	10	5	20
	$D_1[1,1]$	$D_2[2,1]$	$D_5[3,1]$	$D_3[4,1]$	$D_4[5,1]$.

now convert these to cumulative totals

$A_j[\mu,i]$	20	40	50	55	75.

The cash flow from origin zone 1 to zone 3 (j=3) which is 4th nearest to i (μ=4) is

$S_{ij}[\mu,i] = S_{13}[4,1] = 10 \times 10 \times (\exp(-\hat{L} \times 50) - \exp(-\hat{L} \times 55))$,

if L = 0.001 then $S_{13}[4,1]$ = .474
0.0002 = .099
0.00002 = .010

III CALIBRATION

(i) Data Requirements

The shopping model was derived from trip distribution models simply by replacing trips with cash flows. An immediate data problem concerns how we derive estimates of the actual inter-zone cash flows so that we can calibrate the model. Calibration is the process of providing estimates of the unknown parameters we have identified as the independent variables of the model. In the worked examples the effect of different values of $\hat{b}$ was to change the predictions. What we must now do is to devise a value for $\hat{b}$ such that the model provides cash flow patterns which closely correspond to what we would observe in the real world. Unfortunately the only published information is in the form of retail sales in each Local Authority although data for more localised areas may be purchased for 1971; (Census of Distribution: 1951, 1961, 1966, 1971). Thus the model is usually summed over i to provide estimates of zonal sales rather than of the individual flows between zones,

$$\sum_{i=1}^{m} Sij(\hat{a}, \hat{b}, Ai) = \text{sales in zone } j.$$

The remaining fixed variables in the model usually present far fewer problems. Estimates of average retail expenditures can be derived from the Family Expenditure Survey and various estimates of national trends. Population totals can be obtained from the census, annual population estimates, and possibly, the register of electors. The shopping centre attraction term may be represented by a variety of indices of centrality and shopping centre size (Davies, 1970; Kilsby et al., 1973).

The costs of travelling from each zone to every other zone can be estimated by representing each zone as a node on a road network, and then finding the minimum cost or time path through the network. This may sometimes involve trip end penalties for parking, junction congestion, speed limits, link capacities, and so on. Many geographers tend to use straight time distances for expediency but this provides a poor description of costs making it impossible to accurately assess the effects of new road schemes. Other problems relate to the choice of zones especially as there is increasing evidence to suggest that some zoning schemes work better than others. Furthermore, present models work best at a regional scale and tend to break down when applied to individual urban areas.

(ii) Methods of Calibration

To complete the model we have to provide estimates of the parameters $\hat{a}$ and $\hat{b}$. Precise values for these parameters are not usually known in advance and like the parameters in a regression model they have to be estimated in an optimal and consistent manner. However, the parameters in the shopping model are non-linear and cannot be linearised in any simple manner, for example, by taking logs. Early users of these shopping models were unable to calibrate their models in any satisfactory manner and often had to resort to the use of "correction" or "socio-economic" factors to obtain a perfect fit (Murray and Kennedy, 1971; Gibson and Pullen, 1972). Some still resort to these practices. It is, however, most important that good estimates are made of the parameters and that the errors of the calibration process are retained and used to provide confidence intervals on the parameters and the model results.

At present two different approaches are used to provide estimates of the parameters: the method of maximum-likelihood and non-linear least squares.

(iii) The Method of Maximum-Likelihood

The method of maximum-likelihood is based on finding the maximum of a likelihood function which expresses the probability of different parameter values producing an observed result. This method has a strong intuitive appeal and according to it, we estimate the true parameters by any set of parameters which belong to the most plausible set. Often there is a unique maximising parameter set which is the most plausible and thus the maximum-likelihood estimate (Silvey, 1970).

As it happens the maximum-likelihood and entropy-maximising methods are equivalent in this case and the maximum-likelihood estimates for $\hat{a}$ and $\hat{b}$ are those which satisfy the constraints which were previously identified as macrostate system constraints, that is when

$$\sum_{i=1}^{m} \sum_{j=1}^{m} Sij(\hat{a}, \hat{b}, Ai) \times cij - \sum_{i=1}^{m} \sum_{j=1}^{m} Nij \times cij = 0 \quad \text{(cash flow travel cost)} \tag{12}$$

$$\sum_{i=1}^{m} \sum_{j=1}^{m} Sij(\hat{a}, \hat{b}, Ai) \times \ln Wj - \sum_{i=1}^{m} \sum_{j=1}^{m} Nij \times \ln Wj = 0 \quad \text{(cash flow benefit)} \tag{13}$$

where $Sij(\hat{a}, \hat{b}, Ai) = ei \times Pi \times Wj^{\hat{a}} \times Ai(\hat{a}, \hat{b}) \times \exp(-\hat{b} \times cij)$

$Ai(\hat{a}, \hat{b}) = \left(\sum_{j=1}^{m} Wj^{\hat{a}} \times \exp(-\hat{b} \times cij)\right)^{-1}$, and Nij is the actual cash flow from i to j.

Further details of this method of calibration are given by Hyman (1969), Evans (1971), Batty and Mackie (1972), and Williams (1974). You should note that while each parameter has a constraint equation associated with it and that this equation provides a measure of goodness-of-fit, there is not overall indication of how well models with their parameters estimated in this fashion fit the data.

The best estimates of $\hat{a}$ and $\hat{b}$ are those which produce values of Sij() such that these two constraints are satisfied. One way of doing this is to convert equations (12) and (13) into a function of the form

$$F(\hat{a}, \hat{b}) = \frac{(\sum\sum Sij() \times cij - \sum\sum Nij \times cij)^2}{K} + \frac{(\sum\sum Sij() \times \ln Wj - \sum\sum Nij \times \ln Wj)^2}{K},$$

$$\text{where} \quad K = \sum_{i=1}^{m} \sum_{j=1}^{m} Nij. \tag{14}$$

The effect of dividing by K is to convert the cash flow travel cost constraint (equation 12) into a mean cash flow travel cost statistic, and the cash flow benefit constraint (equation 13) into a mean cash flow benefit statistic. Clearly when $F(\hat{a}, \hat{b})$ becomes small both constraints will be satisfied. The method of finding parameter values which minimise this function is left until later.

If we look closer at this form of calibration we find that there are two equality constraints and two unknowns, with the result that the values the unknown parameters can take are completely determined by the constraints. This has some important practical consequences since we cannot obtain an observed value for the mean cash flow travel cost statistic from published sources. However, the mean cash flow benefit statistic can be obtained because it only requires turnover data since we can re-write

$\sum_{i=1}^{m} \sum_{j=1}^{m} Nij \times \ln Wj$ as $\sum_{j=1}^{m} (\sum_{i=1}^{m} Nij) \times \ln Wj$, where $\sum_{i=1}^{m} Nij$ is the turnover in centre j.

It is also important to notice that not all the macrostate constraints are explicit in the entropy-maximising model identified earlier. The origin end constraints find an explicit expression in the Ai() parameters, but the other macrostate constraints are only implicit in the values choosen for $\hat{a}$ and $\hat{b}$. For a model to be calibrated as an entropy-maximising one it is necessary to ensure that all the macrostate constraints are satisfied by an appropriate choice of parameter values. The method of maximum-likelihood ensures that this will happen by explicity satisfying these constraints. However, if these macrostate constraints are not explicity satisfied, as is the case if we use non-linear least squares, then strictly speaking we no longer have an entropy-maximising model but a gravity model with a negative exponential cost function.

(iv) Non-Linear Least Squares

The non-linear least squares approach derives estimates of $\hat{a}$ and $\hat{b}$ such that the residual sum of squares of the differences between the observed and predicted cash flows are minimised;

$$\text{minimise} \quad F(\hat{a}, \hat{b}) = \sum_{i=1}^{m} \sum_{j=1}^{m} (Sij(\hat{a}, \hat{b}, Ai) - Nij)^2 \tag{15}$$

However because no observed values of cash flows, Nij, are available this function has to be modified so that it is based on the difference between the actual Nj, and predicted sales $\sum_{i=1}^{m} Sij()$ in a zone;

$$\text{minimise } F(\hat{a}, \hat{b}) = \sum_{j=1}^{m} (\sum_{i=1}^{m} Sij(\hat{a}, \hat{b}, Ai) - Nj)^2. \tag{16}$$

This can also be expressed as a residual standard deviation, $\hat{\sigma}$,

$$\hat{\sigma} = \text{SQRT}(F(\hat{a}, \hat{b}) /M - (\sum_{j=1}^{m} (\sum_{i=1}^{m} Sij - Nj) /M)^2). \qquad (17)$$

The residual standard deviation is a useful measure of how well the model performs since we would expect the residuals to be normally distributed with a mean of zero. An estimate of the standard deviation is given in equation (17). Thus we could expect about 68 per cent of the residuals to be within plus or minus one residual standard deviation of the model predictions and 95 per cent within two standard deviations.

The non-linear least squares estimates of $\hat{a}$ and $\hat{b}$ should also be unbiased estimates of the maximum - likelihood parameters, although it appears that the maximum - likelihood estimates are nearly always biased. However calibrating the model by this method will not ensure that the entropy-maximsing constraints in equations (12) and (13) are satisfied.

(v) Unconstrained Optimisation of a Function

Both methods of calibration involve finding the minimum of functions which represent how well the model performs. Thus if we only have one parameter, z, and wish to minimise some function of z, say f(z), we could draw a graph of the function for a large number of values of z; see Figure 2. Clearly the smallest value of the function will occur at the bottom of the valley. This particular function is convex in shape and only has one minimum point for the values of z which lie between 1 and 100. However it is possible that had we examined a wider range of values that other minima would occur.
In Figure 3 there are three minima when z ranges from 1 to 200 and the function in only locally convex.

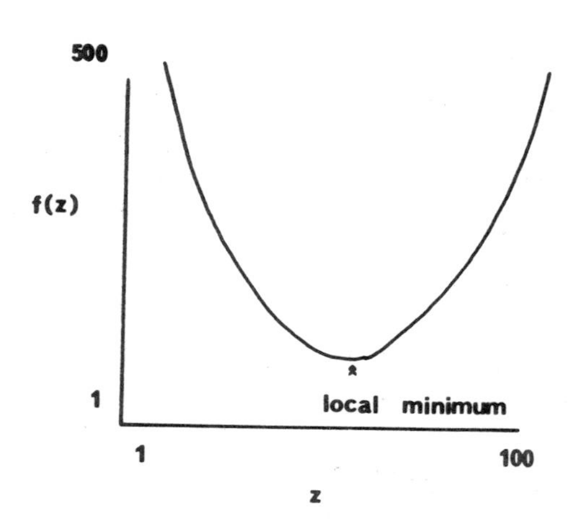

Fig 2. Convex function of one variable, f(z).

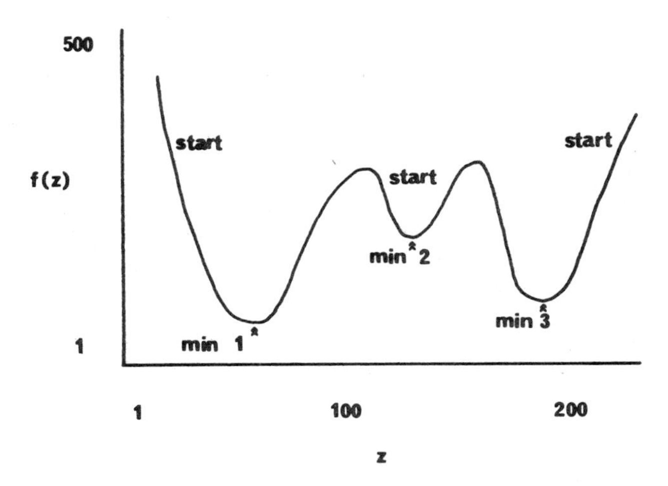

Fig 3. Locally convex function of one variable, f(z).

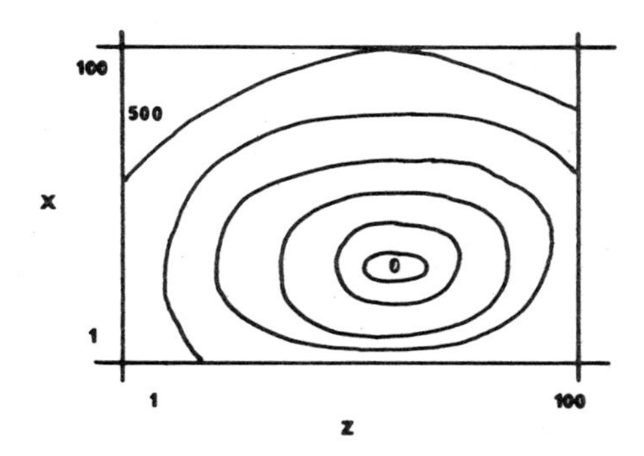

Fig 4. Convex functions of two variables, f(z,x).

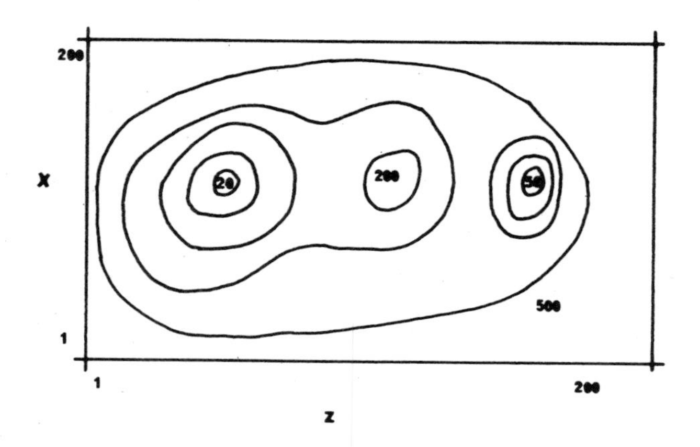

Fig 5. Locally convex function of two variables, f(z,x).

If instead of one parameter we had two, then the function would become f(z,x) which we could draw as a surface. Figures 4 and 5 are surfaces from which Figures 2 and 3 were derived.

Obviously it is not practicable to tabulate a continuous function for all possible real numbers that could be assigned to z and x. Nor is it really satisfactory to select a grid of values since we may miss one or more minimum points. It is far more efficient to use calculus, or a numerical search technique if the function is non-linear. We can calculate the gradient at any point on the curve or surface by deriving an expression for the derivatives or partial derivatives. At each minimum point these terms would be zero. If the function is linear in its parameters we could find the minimum by equating the derivatives to zero and solving a system of linear simultaneous equations. However, our calibration functions are non-linear so we have to use a search technique. These methods use estimates of the gradient to define where to look for the minimum of a function. There is now a wide selection of suitable techniques available Dixon (1972).

These search methods would soon find the minimum point of the function in Figure 2, but in Figure 3 three different minima would be identified according to which local convexity the starting position was in. This situation has important implications for the shopping model since there are restrictions on the range of possible values that can be assigned to the parameters, because the exponential terms are only continuous over limited ranges and the calibration functions are unlikely to be globally convex. These problems can be reduced by exploring the region around a local minimum and by re-starting the search procedure from different positions. It is also very useful to plot the function itself over a fine grid of values in order to be able to identify peculiarities.

IV MODEL PERFORMANCE

(i) Performance During Calibration

So far discussion has focussed on the theoretical aspects of calibration. The problems become far clearer once the behaviour of the model is examined in an empirical context. For this purpose a set of 63 zones covering North-East England is used to demonstrate the model. The variables (sales in a zone W_j, expenditure per person e_i, population in a zone P_i, and car times between zones c_{ij}) all relate to 1961. A distinction was made between food and non-food sales, and sales are used as a measure of attraction.

A non-linear least squares calibration procedure was used to calibrate the entropy maximising model. As a preliminary step a series of response surfaces were generated by plotting the values of one residual standard deviation for a range of values for $\hat{a}$ and $\hat{b}$, see Figures 6 and 7.

For the non-food data the model headed for the so-called bogus or trivial solution, and if you look at Figure 6 showing the model's response surface, you will notice that the minimum of the function occurs when $\hat{a} = 1.0$ and $\hat{b} = 0$ and that the function is convex throughout the domains examined. At this point the model will predict the sales in zone j as equal to

$$\frac{W_j}{\sum_k W_k} \times \sum_i e_i \times P_i$$

and when $\sum_{i=1}^{m} e_i \times P_i = \sum_{j=1}^{m} W_j$, the sales in zone j are identical with the attraction of zone j. Of course the corresponding residual sum of squares is zero at this point. However this solution cannot be considered satisfactory because the model does not now describe consumer behaviour. Several previous studies have in fact approached periliously close to this bogus solution (Nedo, 1970). However, even when we replace sales by a measure of centrality, floorspace, or some other proxy there is still a trivial solution, although it is not a precise one in the sense of a residual sum of squares which approaches zero.

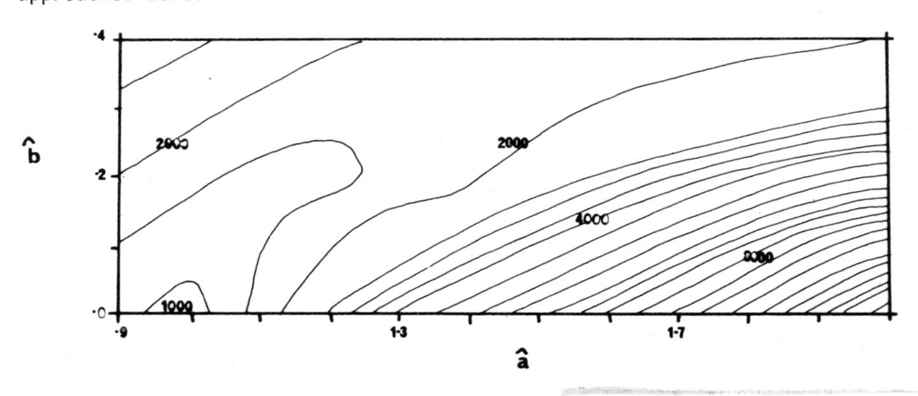

Fig 6. Response surface for non-food data (in £ '000).

The parameters at a hidden trivial solution can be assigned almost any value which represents the equivalent of regressing the attraction term on turnover. This is also unacceptable and is a direct result of not having an observed matrix of cash flows.

The food data produces some surprises since the model sometimes found a non-trivial solution depending on the initial starting values of the parameters. The negative exponential model's parameters at this point are $\hat{a}$ = 1.140 and $\hat{b}$ = 0.169 with a residual standard deviation of £838 000. Figure 7 describes the response surface and it is most noticeable that the model has a local convexity in the region labelled A. Once the minimisation routine reaches this area it is unable to find the trivial solution and converges to a local minimum within it. If the starting values for the parameters are such that the search avoids this critical area, the trivial solution is found. In cross section this situation is similar to the graph in Figure 3. Clearly this non-trivial solution is data dependent and not acceptable.

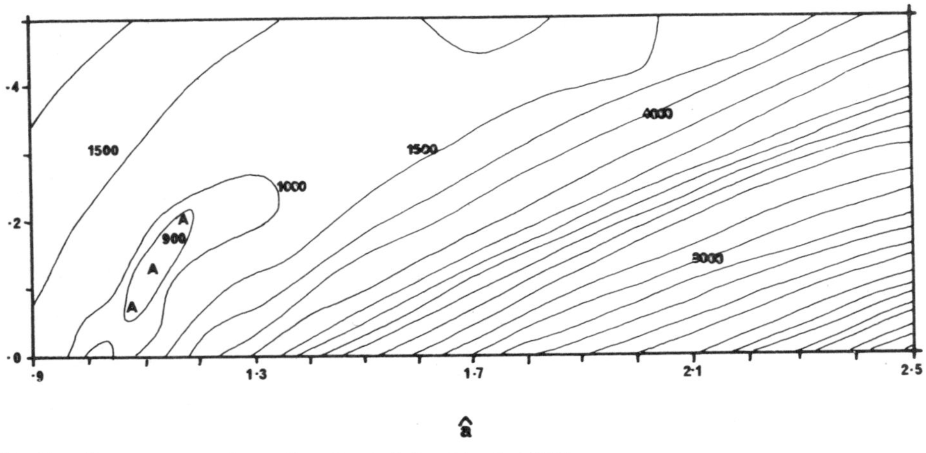

Fig 7. Response surface for food data (in £ '000).

Before leaving least squares calibration, it is of some interest to note that the one parameter models of Batty and Saether (1972) are equivalent to drawing a line perpendicular to one of the axes of the response surfaces to represent the value of the parameter which is assumed to be held constant. For example, if in Figure 7, $\hat{a}$ is set to 1.7 and $\hat{b}$ allowed to vary, then by erecting a vertical line to represent $\hat{a}$ = 1.7 the value of $\hat{b}$ would be approximately .39 and the residual standard deviation about £1.4M. The use of a constrained optimisation procedure would have a similar effect with the parameter values clinging to the boundaries of the feasible zone closest to the best solution. To summarise, then, it is impossible to obtain a satisfactory calibration of the model using non-linear least squares when cash flow data is not available.

Calibration of the entropy-maximising model by the method of maximum-likelihood will not find a trivial solution but without cash flow information the improvement in objectivity is largely illusionary. The parameter values are entirely dependent on the user being able to provide an observed value for the mean cash flow travel cost statistic in equation (12). Without detailed observed cash flow information this statistic can only be guessed at. The mean cash flow benefit statistic can be estimated from the available data. It is also important to note that we cannot estimate $\hat{a}$ and $\hat{b}$ independent of each other since they are related through the Ai ($\hat{a}$, $\hat{b}$) term. Most studies using this method guess at a value for the mean cash flow travel cost statistic, or carry out a very small survey to provide an estimate of it (Smith, 1973; Mackett, 1973; Wade, 1973).

Another way of determining an appropriate value for the mean cash flow travel cost statistic is to explore the goodness - of-fit of the model over a wide range of values. Table 1 shows the result of examining a range of different values for the food data.

TABLE 1

FOOD SALES: RESIDUAL STANDARD DEVIATIONS FOR VARIOUS TRAVEL COSTS

Mean cash flow travel cost (in minutes)	$\hat{a}$	$\hat{b}$	Residual standard deviations (£'000)
9	1.537	.4348	1214
12	1.119	.1906	887
14	1.063	.1431	889
19	1.021	.0844	965
25	1.015	.0491	T053
29	1.015	.0327	1004
35	1.013	.0131	612
39	1.008	.0015	122
43	1.001	.0095	609

A local convexity in the residual standard deviation exits between 12 and 14 minutes and this is in fact equivalent to the non-trivial solution identified using non-linear least squares calibration of the negative exponential model. A similar table was produced for the non-food data but displayed no local convexity confirming the results of Figure 6. Furthermore, it is only when the parameters approach very close to the trivial solution values of 1.0 and 0.0 that the residual standard deviations become small enough to be acceptable. For the food data the trivial solution has a mean cash flow travel cost of 39.564 minutes. The other statistic is already at the trivial solution value. It is apparent, therefore, the maximum - likelihood calibration of the entropy-maximising shopping model also suffers from fundamental problems when adequate cash flow information is not available.

(ii) Evaluation of Model Goodness-of-Fit

Despite severe problems with calibration a large number of planning studies illustrate the fact that the applied use of these models continues. It is useful then to evaluate the goodness-of-fit of a selection of the models which could be calibrated, albeit with misgivings.

The residuals produced by the two non-trivial food models calibrated using non-linear least squares and the method of maximum-likelihood, as well as three non-food models with assumed mean cash flow travel costs of 20, 25 and 30 minutes, were selected for an assessment of model performance. It is very difficult to define generally acceptable levels of performance although there are some significant statistical indicators. In particular it is desireable that the residuals are not autocorrelated when ranked by possible deflator variables, that the residuals are homoscedastic, that the model's parameters are not highly correlated, and that the overall level of fit is acceptable. Violations of these diagnostics will provide a guide to possible sources of error in model design as well as provide an indication of how useful the model actually is in an empirical situation.

The residuals were ranked by size of centre on the assumption that the magnitude and sign of the residuals were affected by size (a deflator variable), but according to a two-tailed runs test the pattern of residual signs were all random at a 5 per cent confidence level. However, the residuals did show considerable heteroscedasticity (that is the variance of the residuals was not the same wher calculated for various subsets of centres) at less than a 1 per cent level, using a method devised by Goldfeld and Quandt (1965). The residuals also displayed very significant positive spatial autocorrelation according to a Cliff and Ord (1973) statistic. This test provides a measure of the presence of spatial pattern in the residuals when in theory there should be none. Finally, it was observed that none of the parameters were highly intercorrelated and all the standard errors were small.

The size of the residuals relative to the size of the shopping centre concerned was also examined. If we decide that the model must be able to describe the pattern of sales such that one residual standard deviation is less than 25 per cent of the sales in a centre before it can be regarded as being acceptable, then only 19 centres would qualify for the food data and 11 for the non-food data out of a total of 57 centres. Yet this assumes an approximate 90 per cent confidence limit of plus or minus fifty per cent of the sales in a centre. If this critical threshold is reduced to 25 per cent then only 8 food centres and 9 non-food centres would qualify. It is apparent, therefore, that very few of the shopping centres can be adequately represented by the existing models.

These results confirm the findings of an earlier study using total sales, Openshaw (1973). In this study it was found that the best performance resulted from using a single parameter Lakshmann-Hansen model with $\hat{a} = 1.1$ (fixed) and $\hat{b} = 2.59132$, and $\hat{b} = .19245$ for the negative exponential model. The resulting standard deviations were £1.769M and £1.664M respectively. An intervening opportunity model was also tried with the pattern of opportunities being represented by sales which were ranked by distance from each origin zone. This model has no trivial solution but provided a very poor description with a residual standard deviation of £7.5M. This is not really representative of the behaviour of an intervening opportunity model since studies with a journey-to-work version of the same model produced results which were only about 10 per cent poorer than the norm.

The importance of these poor overall levels of performance assumes greatest significance in an applied situation. Disaggregation of the model only seems to increase the level of the relative errors. A geographer interested in model description may find it far easier to absorb errors of plus or minus 50 per cent than a planner, although on previous experience neither have taken much notice of model performance. Others have deluded themselves that a high correlation coefficient means that all is well, whereas more sensitive measures of goodness of fit present a very different picture.

V THE DATA PROBLEM

(i) Shopping Trip and Cash Flow Data

It was noted earlier that many of the calibration problems result from the absence of observed trip and cash flow information. It is necessary, therefore, to consider collecting this data using a form of diary survey. A diary survey involves households in each of the model zones keeping a record of all their shopping trips over a period of at least one week, with details of places visited, goods bought, mode of travel, origin zone, etc. Several diaries have been collected (Davies, 1973; Bruce and Daws, 1971; Rodgers, 1974) but very few have been designed primarily to provide information for calibrating a shopping model (Berkshire County Council, 1974). It is vital that the sample size is sufficient to ensure an adequate description within acceptable sampling errors. The Berkshire shopping diary has an estimated completed size of about 6000 households.

Conventional sample size theory does not prove very useful when applied to trip distributions as distinct from trip end totals. It seems likely that for a system of 100 zones a sample of between 4000 and 6000 households in each zone would be necessary if the sampling errors on the trips or cash flows between the zones are to be kept within the range of plus or minus 10 per cent at a 90 per cent confidence limit. A far smaller sample size may be adequate if a mean cash flow travel cost statistic is to be estimated but unless the cash flows are also known then it will be impossible to verify how well the model actually performs during calibration. Nevertheless, if only a small sample of households is available then maximum-likelihood calibration offers greatest utility. In this respect the availability of the County Surveyor Society's Trip Rate Data Bank provides a source of mean trip length statistics at a national level, provided sampling errors are ignored.

An alternative approach is to adopt a far simpler kind of survey questionnaire and record details of those centres which a household "normally" visit for various goods. A survey along these lines was carried out by Durham County Council in 1974. This information makes it possible to consider constraining the model so that large numbers of $S_{ij}()$ elements are zero whereas the conventional model usually allocates some expenditure from every zone. Quite different sets of $S_{ij}()$ elements could be set to zero for different types of goods. This is quite useful since a large proportion of the cash flows described by the model may be regarded as very unlikely to ever occur and by constraining these trips to zero it is possible to place severer restrictions on consumer behaviour that at present is allowed.

(ii) Estimating the Missing Cash Flows

It may also be possible to provide estimates of the cash flows themselves using a conditional probability estimation technique based on the same entropy-maximising approach used to derive the shopping model. Chilton and Poet (1973) have shown how it is possible to use entropy-maximising techniques to expand abbreviated census data and this approach can be adapted for estimating shopping trip cash flows. The essence of the problem is to derive estimates of a matrix of cash flows when our information is limited to the trip end totals e_i, P_i, and W_j. However, for there to be an unique solution a large number of the cash flows must be known. So some form of sample survey is still needed to provide estimates of some of the flows or at least indicate which flows can be set to zero. There are several variations of data gap filling techniques and hybrid calibration procedures that could be used.

Clearly, data estimation can never be a substitute for survey information, but it is becoming apparent that the resources necessary for a large scale survey may not be available and some kind of data estimation or data enhancement technique will be required. The fact that many shopping patterns are at least regional in scale does little to encourage the co-ordinated collection of this data even though it is a pre-requisite for building empirically acceptable shopping models.

VI ASPECTS OF MODEL DESIGN

(i) Some Consequences of Entropy-Maximising

The major characteristics of the model result from the entropy-maximising deriviation of it. It is very important to note the importance of the macro-state constraints both on the design of the model and on its subsequent performance. The model assumes a closed system in an equilibrium state such that the values of these constraints will be fixed regardless of the actual distribution of cash flows. Thus if the distribution of cash flows was such that short distance trips predominated we would expect the mean cash flow travel cost statistic to be reduced, and conversely, if long distance trips predominated that it should increase. However, in practice the model seeks to compensate an increase in some long distance trips by decreasing other trips so that the constraints are preserved.

It is apparent, therefore, that these macrostate constraints have profound behavioural implications. It could well be desireable that these constraints should be expressed as inequalities, for example,

$$\sum_{i=1}^{m} \sum_{j=1}^{m} Sij(\hat{a},\hat{b},Ai) \times cij \leqq \sum_{i=1}^{m} \sum_{j=1}^{m} Nij \times cij,$$

or as bounds specifying upper and lower limits

$$\text{upper value} \geqq \sum_{i=1}^{m} \sum_{j=1}^{m} Sij(\hat{a},\hat{b},Ai) \times cij \geqq \text{lower value},$$

rather than as precise equalities. This would greatly reduce the significance of these constraints on some important aspects of the model. However, an entropy-maximising model could not handle these constraints and one must suspect that part of the reason for the macrostate equality contraints is mathematical convenience.

(ii) Impact Studies

Another practical consequence of these constraints occurs when the model is used for an impact study. An impact study usually takes the form of running the shopping model in such a way as to assess the impact of additional shopping provision in a simulated present or future situation. The model is run first without the proposed additional facilities and then again with the inputs modified so as to reflect the proposed changes. The differences in sales attributed to each zone has been interpreted as a measure of impact. However, the existence of these macrostate constraints ensures that the shock of additional facilities is spread throughout the system in an attempt to preserve the trip end constraints and keep the other two constraints at their previous levels. It is also necessary to make some assumptions about what effect the new facilities might have on the models parameters. This is discussed in detail in section (iii).

In impact studies some major problems result from the effects of ensuring the trip end constraints remain satisfied. This can be demonstrated by reference to a case study. In 1971 a 100 000 sq. feet hypermarket proposal was examined by Durham County Council. A forecast was made of a simulated 1981 situation and the hypermarket was estimated to receive a turnover of £5.5M with every shopping centre in the set of 64 zones experiencing a reduction in turnover. If the trade area of the hypermarket was restricted to 30 minutes of car travel time, then the total sales loss in this area only amounted to £2.8M, or about half the total allocated by the model to the hypermarket. It is apparent then that the indirect effects of the hypermarket spread far beyond the maximum likely trade area partly as a result of the model maintaining the trip end constraints. In fact when the gross effects were calculated by preventing the model re-established its proper equilibrium state, sales losses were found to be between two and three times the net amount in many of the zones within the 30 minute trade area.

(iii) Conditional Forecasts of Future Levels of Sales

The conventional method of deriving forecasts like impact studies is to change some or all of the independent variables in the model in order to simulate a future state. The first set of problems concern the need to take into account forecast errors in the independent variables ei, Pi, Wj, and cij. Precise forecasts cannot be made of any of these variables. Many of these forecasts describe at best the mean of a distribution of possible values and provided their distributions are known monte-carlo techniques could be used to generate sets of alternative equi-probable forecasts. Re-running the model with a large number of different sets of forecasts will provide an estimate of forecast variability. Likewise, it is possible to assess the effect on the forecasts resulting from the standard error of the parameters and the standard deviation of the residuals.

Other problems relate to the assumptions made about the future values for $\hat{a}$ and $\hat{b}$. Furthermore, if we accept an entropy-maximising model, then the forecasts obtained must be consistent with the assumptions of this approach. Let us consider an actual forecast situation. The data relates to the previous set of 63 zones but this time we will use total sales. The method of maximum-likelihood was used to provide estimates of the parameters of an entropy maximising shopping model in 1961. An assumed mean cash flow travel cost of 16 minutes and an observed mean cash flow benefit of 9.97839 were used. Estimates of $\hat{a}$ and $\hat{b}$ were derived such that the largest difference between the actual and predicted values of these constraints was less than 0.1×10^{-13}, the value of the associated function (equation 14) was $0.2753124 \times 10^{-27}$, and parameter values of $\hat{a} = 1.0366$ and $\hat{b} = 0.12400$ were obtained. Simply by substituting estimates of ei, Pi, Wj and cij in say 1981, we can derive some forecasts of turnover levels; see column (a) in Table 2. Since we have retained the calibration values of $\hat{a}$ and $\hat{b}$ we have assumed that the 1981 mean cash flow travel cost and mean cash flow benefit statistics are still at their 1961 values.

However, as we have seen the constraints necessary to preserve these values are explicit only in the calibration procedure and not in the model. So we find that our 1981 forecasts no longer satisfy these constraints and have a mean cash flow travel cost statistic of 15.8967 and a mean cash flow benefit statistic of 9.8254. In other words we no longer have a model consistent with an entropy-maximising derivation. Many users seem to have fallen into this trap, indeed most of the literature on entropy-maximising spatial interaction models relates to model derivation rather than to forecast applications.

A rather different situation arises when we know about this problem and wish to ensure that the 1961 macrostate constraints are preserved in our 1981 forecasts. There are now two possible approaches. We can simple re-calibrate the model for 1981 and derive new estimates of $\hat{a}$ and $\hat{b}$ such that these constraints are satisfied, but we would no longer have any real measure of how well the model performs and are unable to estimate confience intervals for it. Accordingly, we re-run the maximum-likelihood procedure and derive new estimates of $\hat{a}$ = 1.17906 and $\hat{b}$ = .126538 and a value for the calibration function of $.451987 \times 10^{-18}$. However these small changes in parameters produce a dramatic effect on the forecasts we obtain; see Table 2 column (b). Alternatively, we can retain the 1961 parameter estimates and seek to relax the trip end constraints so that the mean cash flow travel cost and mean cash flow benefit constraints are retained by reformulating the model as a non-linear programming problem (see Section VI(v)). But depending on the degree of relaxation we may no longer have an entropy-maximising forecast.

Another kind of situation would arise if we could derive estimates for these two statistics and then re-calibrate the model. However it is not likely that these statistics can be forecast exogenously in any realistic manner. Table 2 column (c) shows the forecasts produced if we assumed a 1981 mean cash flow cost of 12 minutes and mean cash flow benefit of 8, which has $\hat{a}$ = 1.43727 and $\hat{b}$ = .231983.

TABLE 2

1981 TURNOVER FORECASTS ('000 POUNDS)

Zone	(a) 1961 parameters retained	(b) Re-calibrated 1981 parameters	(c) Re-calibrated 1981 parameters
1	107600	120206	119098
6	20881	19074	16787
9	5413	4151	3491
38	3980	3216	3607
41	2267	2104	2507
51	1880	1523	1347
53	37	17	6
60	20	10	6

If the model has been calibrated using non-linear least squares then we can derive forecasts by holding the parameters constant since we would not know in which direction to change them, although we can not be too happy with the situation. In fact even if we can obtain estimates of our variables (ei,Pi,

Wj, cij) for 1981 with only an error of about 25 per cent, then the differences obtained by simulating this forecast error using monte-carlo methods is huge. If we add the problems of forecasting in a manner consistent with the entropy-maximising approach the situation becomes even more precarious. Finally, if we add a measure of model performance error, the situation becomes intolerable. In the short term it is likely that it is only in the field of model performance in a descriptive situation that any improvements can be expected.

(iv) Some Improvements to Model Design

One of the attractions of mathematical modelling lies in the ability of the investigator to manipulate the structure of a model to accomodate new ideas, and by testing the model prove or disprove the underlying theory. Certainly at present insufficient attention has been given to the empirical testing of the theory based models that are in use to-day. To illustrate this process of empirical testing and model improvement attention is drawn to two additional weaknesses in the current model.

A major source of dissatisfaction with the existing shopping model in an impact study concerns the continuous probability surface assumed by the model. Provided Wj is greater than zero then the probability of a flow of money travelling from origin zone i residents to destination zone j shops is

$$\frac{Wj^{\hat{a}} \times \exp(-\hat{b} \times cij)}{\sum_{k=1}^{m} Wk^{\hat{a}} \times \exp(-\hat{b} \times cik)} > 0$$

whereas in reality it is likely to be spatially discontinuous. Indeed it is possible that the behavioural concepts on which the model is apparently based, allows the consumer too much spatial freedom. This is a reflection of both the nature of the cost or deterrence function and of the competition or balancing terms in the model. This is of course in accordance with the entropy-maximising model's principle of maximum freedom within the external macrostate constraints. Nevertheless, the far more restrictive postulates of central place theory may provide a more satisfactory alternative. It may be necessary, therefore, to design a model in which local trips are restricted to a few competing centres with unsatisfied consumer moving to successively higher order centres.

The PRAG hierarchical shopping model provides one way of restricting consumer behaviour (Wade, 1973). A study area is sub-divided into a set of sub-systems each of which could represent different levels of the central place hierarchy or areas of functional unity. Different model parameters could be provided for each of these sub-systems. However, this approach presupposes that the various levels of the central place hierarchy can be identified on the ground for each level of disaggregation considered. The major advantage seems to be in terms of computing efficiency at a time when this is no longer a major practical consideration.

A far more useful and general procedure has been developed by Openshaw and Connolly (1976). They have developed a completely general deterrence function with one or more discontinuities in it. Pairs of trips are classified to one of several functions with the form of the function changing dynamically to match the data allocated to it. The trip pair classification could be based on central place concepts, on notions of distance related changes, or on macro-geographical structure. This amounts to a recognition of the existence of spatial discontinuities in the probability surface of the model. With the discontinuities being structured to match the unique spatial structure of a study area, while the form of the function adjusts to match whatever fits the data in a specific trip set best. The existing one-fixed-function no-discontinuity model is therefore a very special case of this more general approach. Such a model is very geographical since it can explicitly reflect geographical structure, but requires detailed trip information before it can become operational in a shopping model context.

Another major problem concerns the mis-interpretation of the spatial interaction model in a retail context. The existing models are essentially models of shopping trips but they are used as models of cash flows. However, the conversion of a trip pattern into cash flows is made on the basis that each trip is weighted equally in terms of expenditure and that the cash flows are directly proportional to trip frequency. In reality each trip will not carry with it the same expenditure and the conversion of trips to cash flows may be expected to be related to factors which are quite independent of those which influenced the pattern of trips. The current generation of shopping models only provide a description of the first stage of this process (Openshaw, 1974). It may be necessary to consider modelling the shopping trip distribution first and then to convert the trip pattern to cash flows using a very different type of model.

(v) Alternative Non-Entropy-Maximising Spatial Interaction Models

By adopting an empirical as distinct from a theoretical and mathematical perspective it is possible to develop a wide range of alternative spatial interaction models. Let us re-write the entropy-maximising model as

$$Sij(\hat{a},\hat{b},\hat{A}i) = ei \times Pi \times \hat{A}i \times Wj^{\hat{a}} \times \exp(-\hat{b} \times cij) \quad (21)$$

The $\hat{A}i$ term is now a parameter and not the usual balancing constraint. We could calibrate this unconstrained model using non-linear least squares by finding the minimum of

$$F(\hat{a},\hat{b},\hat{A}) = \sum_{i=1}^{m} \sum_{j=1}^{m} (Sij(\hat{a},\hat{b},\hat{A}i) - Nij)^{2}. \quad (22)$$

This model would result in a far smaller sum of squares of residuals than any of the singly constrained models since we pay for the constraints in terms of an increased sums of squares. To convert this model into a singly constrained model, introduce m equality constraints of the form

$$\sum_{j=1}^{m} Sij(\hat{a},\hat{b},\hat{A}i) - ei \times Pi = 0$$

This model has m+2 parameters and m non-linear equality constraints and would produce results identical to a negative exponential model. If we add two more equality constraints of the form

$$\sum_{i=1}^{m}\sum_{j=1}^{m} Sij(\hat{a},\hat{b},\hat{A}i) \times cij - \sum_{i=1}^{m}\sum_{j=1}^{m} Nij \times cij = 0$$

$$\sum_{i=1}^{m}\sum_{j=1}^{m} Sij(\hat{a},\hat{b},\hat{A}i) \times \ln Wj - \sum_{i=1}^{m}\sum_{j=1}^{m} Nij \times \ln Wj = 0$$

and minimise the residual sum of squares function subject to these m+2 constraints we would derive estimates of the parameters, $(\hat{a},\hat{b},\hat{A}i)$, very similar to those produced by maximum-likelihood calibration of the entropy-maximising model. By re-considering the model in such a way that all the constraints are inherent in the calibration procedure, we have a considerable degree of freedom to experiment empirically by adding additional constraints, introducing inequality constraints, and changing the form of the function.

There has been a tendency to think almost exclusively in terms of the entropy-maximising model, yet there are a large number of alternative models available which have yet to be investigated. Indeed, if you adopt an empirically orientated perspective there is far more to be gained by developing model forms which work regardless of the quality of their theoretical derivation. The entropy-maximising approach offers very significant short term advantages but it is becoming obvious that for practical applications the long term solutions may lie elsewhere.

(vi) Constrained Optimisation of Non-Linear Functions

The previous discussion of function minimisation assumed that the parameters were not constrained in any way and that they could, in theory, be any real number. It is also possible to minimise a function so that the parameters are constrained to a certain range of numbers. Suppose we wish to constrain the values parameter z can take in Figures 2 and 4. The simplest type of constraint is an inequality constraint, for example, that $z \leq 30$ or z must be less than or equal to 30. This constraint restricts the minimum we can find to the region indicated in Figure 8. A far more committing form of constraint is provided by an equality constraint, for example, than z=30. The feasible region is now restricted to a straight line which represents z=30 and a minimum of the function occurs where the function intersects this line. If the function only has one parameter then the only feasible minimum occurs when z=30. If we have two parameters and two constraints, z=30 and x=50, then the feasible region is the point of intersection of the lines representing these constraints, and thus the constrained minimum of the function. This is essentially what happens with maximum-likelihood calibration. These constraints can also be non-linear, for example,

$$z^2 + 2x^2 = 70$$

and although the problem is now far more complex, a solution can often be found using recently developed techniques. Thus it is possible to provide solutions to the non-linear programming problems described in Sections VI (v) and VII.

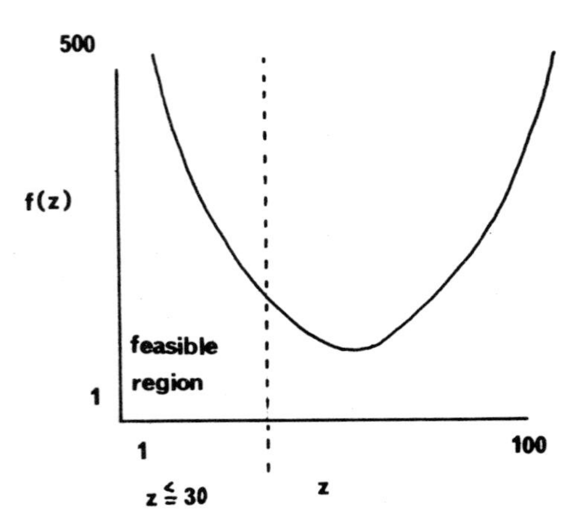

Fig 8. Constrained optimisation.

VII SHOPPING LOCATION-ALLOCATION MODELS

(i) An Elementary Model

To conclude this discussion of shopping models it is necessary to look briefly at the early stages of the development of location-allocation versions. These models consist of shopping models structured in such a way as to locate the distribution of population and shops so as to optimise some measure of overall plan performance. A simple aim may be to minimise the total travel cost of the shopping trips by changing the spatial distribution of population and shops subject to constraints in the form of upper and lower bounds on zonal totals. This model can be expressed as:

$$\text{minimise } F(\underset{\sim}{\hat{W}},\underset{\sim}{\hat{P}}) = \sum_{i=1}^{m}\sum_{j=1}^{m} Sij(\hat{W}j,\hat{P}i,Ai) \times cij \qquad (23)$$

where $Sij(\hat{W}j,\hat{P}i,Ai) = ei \times \hat{P}i \times \hat{W}j^{a} \times Ai(\hat{W}j) \times \exp(-b \times cij)$

$$Ai(\hat{W}j) = \left(\sum_{j=1}^{m} \hat{W}j^{a} \times \exp(-b \times cij)\right)^{-1}$$

subject to constraints of the form

$$u\hat{P}i \geqq \hat{P}i \geqq l\hat{P}i$$
$$u\hat{W}j \geqq \hat{W}j \geqq l\hat{W}j$$

and possibly $\underset{\sim}{c}^{T}j\underset{\sim}{\hat{P}} \geqq dj \qquad j=1, \ldots m$

$$\underset{\sim}{c}^{T}j\underset{\sim}{\hat{W}} \geqq dj \qquad j=m+1, \ldots m+m;$$

where $l\hat{P}i$ is the lower bound on population in zone i,
$u\hat{P}i$ is the upper bound on population in zone j,
$l\hat{W}j$ is the lower bound on $\hat{W}j$ in zone j,
$u\hat{W}j$ is the upper bound on $\hat{W}j$ in zone j,
the other constraints specify upper or lower limits or fixed total for $\hat{P}i$ and $\hat{W}j$ in sets of zones (we could also fix limits to the size of these parameters in sub-regions),
$\hat{P}i$ is the population in zone i, $\hat{\underset{\sim}{P}}$ the vector of all populations selected so as to minimise F(), $\hat{W}j$ is the attractiveness, or floorspace, in centre j, also selected so as to minimise F()

As an example of this kind of model, an attempt was made to find the optimum population distribution for the set of 63 zones for food sales, but with the size of shopping centre held constant. The initial cost of these food shopping trips was 3492203 minutes and the smallest possible was 3397750 minutes. A change in mean trip length from 12.4 to 12.0 minutes. The optimum pattern of population was quite different from the initial one yet only a few zones had optimal values on the limits of their bounds which in this case was ± 15 per cent of their original values. The exercise could have been repeated allowing the attraction parameters to change as well as the population ones. It is also possible to permit the a and b terms to vary subject to upper and lower bounds since this can reflect our uncertainty about the precise values they can, or will, have in any situation.

(ii) A Family of Models

More recently Wilson (1974) has suggested two additional objective functions which may also be useful in this context. He defines a measure of global retailer profit as

$$\Pi(\hat{\underset{\sim}{W}}) = \gamma \sum_{i=1}^{m} \sum_{j=1}^{m} Sij(\hat{W}j, Ai) - \sum_{j=1}^{m} pj \times \hat{W}j$$

where γ is the proportion of total sales which is an operating profit,
pj is the rent per square foot of shopping space, and
Wj is the floorspace of the shops in zone j;
and a measure of global consumer welfare as

$$Z(\hat{\underset{\sim}{W}}) = \sum_{i=1}^{m} \sum_{j=1}^{m} Sij(\hat{W}j, Ai) \times (\frac{a}{b} \times \ln Wj - cij) \tag{25}$$

where $\frac{a}{b}$ lnWj represents the size benefit of shopping at j, and
cij the cost or disutily of travel from i to j.

By changing the distribution of shopping floorspace either of these measures can be optimised. By combining these objective functions with the cost optimisation model described earlier it is possible to develop several alternative versions of location-allocation model with different parameter and constraint mixes (Openshaw and Connolly, 1975).

If the aim is to optimise Wilson's consumer welfare function subject to upper and lower limits on total shopping travel costs, retailer profits, population, and shopping floorspace then the model becomes

$$\text{maximise } F(\hat{\underset{\sim}{P}}, \hat{\underset{\sim}{W}}) = \sum_{i=1}^{m} \sum_{j=1}^{m} Sij(\hat{P}i, \hat{W}j, Ai) \times (\frac{a}{b} \times \ln Wj - cij) \tag{26}$$

subject to the constraints:

$$\prod u \geqq \gamma \sum_{i=1}^{m} \sum_{j=1}^{m} Sij(\hat{W}j,Ai) - \sum_{j=1}^{m} pj \times \hat{W}j \geqq \prod 1$$

$$Cu \geqq \sum_{i=1}^{m} \sum_{j=1}^{m} Sij(\hat{W}j,\hat{P}i,Ai) \times cij \geqq Cl$$

$$uWj \geqq \hat{W}j \geqq lWj$$

$$uPi \geqq \hat{P}i \geqq lPi,$$

where $\prod u$ and $\prod l$ are the limits of acceptable profits, and
Cu and Cl are the limits of acceptable total travel costs.

Problems like this can be solved using non-linear programming techniques although the task is easier if the constraints are linear. Nevertheless methods exist now that can solve these problems.

This approach to the development of shopping location-allocation models presents some conflict with the entropy-maximising approach. Here we are sometimes explicitly minimising a macrostate constraint and simultaneously changing other constraints. Certainly this does not fit into the neat relationship Wilson has developed between Linear Programming and entropy-maximising but there is no reason why this is a disadvantage in an empirical sense. In fact models along these lines offer tremendous potential for planning studies. A major difficulty relates to the acceptability of the conventional spatial interaction model as a good trip descriptor. Indeed allocation-location models would be far more useful if the performance of the Sij() model was better. Nevertheless this kind of model can produce very useful results for both the geographer and planner when run as a simulation model in the context of testing alternative hypotheses and policies relating to the spatial distribution of population and shopping activities.

VIII CONCLUSION

(i) Availability of Computer Programs

Most people will learn more about shopping models by running problems for which they have data and which they understand. Interpretation is possible by relating their knowledge about the data to the output from the computer. Many of the empirical problems have to be seen to be really appreciated since this provides the basis for the development of better theory.

There is a wide choice of computer programs available. It is suggested that an automatic calibration routine should always be used, and that at present, it is less hazardous to use maximum-likelihood calibration. Single parameter automatic calibration routines have been published by Baxter (1974) and Mountcastle, Stillwell, and Rees (1974) in the context of general purpose spatial interaction models. A far more comprehensive set of calibration procedures covering various two-parameter models using non-linear least squares, maximum-likelihood, minimax, and chi-square methods is to be provided, Openshaw (1977). No complete set of programs for the location-allocation models are available yet but it is expected that these will be forthcoming during 1977

(ii) Some Broader Aspects of Building Geographical Spatial Interaction Models.

Many of the conclusions derived from the study of shopping models can be applied to other spatial interaction models. It is appropriate then to provide some general conclusions in the way of advice for geographers interested in this field.

The appeal of the entropy-maximising approach has been so great that there is a current tendency to see everything in terms of it, resulting in a profound blinkering effect on some aspects of contemporary research. The hard-line entropy-maximiser probably interprets the structure of spatial interaction models as portrayed in Figure 9a. This monograph has tried to show that perhaps Figure 9b is more appropriate. The planner is most unfortunate in his perspective since he is in the unique position of having to use his models in application studies and is subject to considerable restrictions on the time available to him, see Figure 9c. It is most important that both geographer and planner adopt the correct perspective of these models since this will largely dominate subsequent research and application activities.

A geographer also has to consider in what ways are these models relevant to him as a geographer. Certainly few of the so called new geographical techniques are even remotely geographical, in fact like the generic terms 'spatial interaction model' they justify the adjective 'spatial' or 'geographical' mainly in the narrow sense of describing the nature of the data input. Yet the results of ignoring geographical structure are becoming very clear as empirical studies provide increasing evidence of poor levels of performance. For example, the apparently simple task of defining zonal boundaries is fraught with scale and aggregation problems. Different boundary choices can provide quite different results and a deliberate heuristic boundary selection algorithm can pay dividends in terms of improvements in model performance. Furthermore, the traditional way of interpreting the parameters in a globally determined deterrence function is as a reflection of overall geographical structure. This overlooks the alternative view that geographical structure should itself determine the nature and shape of the deterrence function.

In their search for geographical theory and in the total revolution against uniqueness, geographers seem to have overlooked a basic fact. Theory by definition cannot be unique in a geographical sense, but the application of theory in the form of a model to a study area will involve the description of geographical information which is quite unique. The calibration of a fixed function in the spatial interaction model is the only opportunity whereby the theory, of which the model is an expression, can be adapted to the spatial structure of a study area. Clearly the amount of adaptation possible is restricted by the general form of the function itself, and any interpretation given to the parameters in the function cannot have very much geographical relevance. It is far more sensible to build the relevant aspects of spatial structure into the model in an explicity manner during the calibration process, indeed it has already been shown how this can be achieved. The absence of a rigorous theoretical (mathematical as distinct from geographical) justification for such modification is not of any great practical importance. Certainly the dangers of proceeding along a path of empirical models without the backing of mathematically rigorous theory is far less dangerous than the

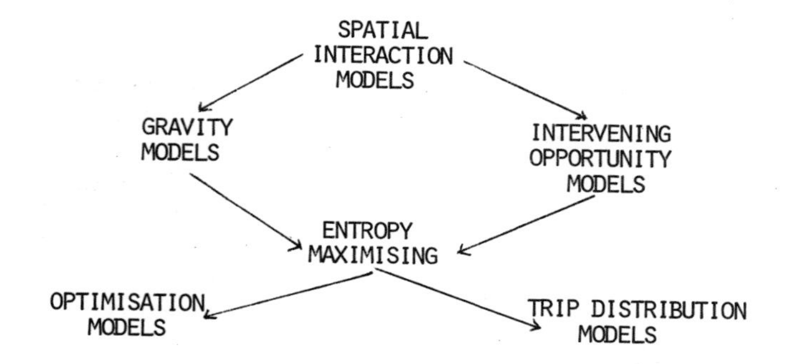

(a). Hard-line entropy maximiser

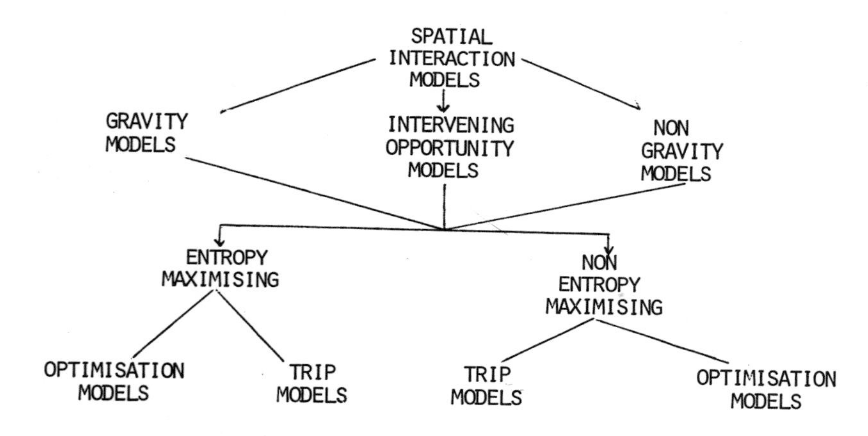

(b). A broader outlook

(c). An operational planner's outlook

Fig 9. Three perspectives on spatial interaction modelling

current habit of adopting a theory based model which lacks geographical significance and which has not been rigorously tested because of data deficiency.

Furthermore, despite the utility of entropy-maximising as a model building technique no geographer should overlook the fact that it is totally aspatial. The microstates have a spatial representation as flows between points, which are in practice aggregated to zones. These zone to zone movement patterns are now regarded as microstates but clearly it is not really valid to assume that each zonal microstate is equiprobable. In fact zones are often of different shapes and sizes and it may be expected that the larger zones may have a greater probability. One solution has been developed by Batty (1974) and involves weighting the microstate probabilities by size of zone. Maximising his definition of spatial entropy would produce a shopping model of the form

$$Sij(\hat{a},\hat{b},Ai) = ei \times Pi \times \Delta\, Xj \times Ai(\hat{a},\hat{b}) \times Wj^{\hat{a}} \times \exp(-\hat{b} \times cij)$$

$$Ai(\hat{a},\hat{b}) = \left(\sum_{j=1}^{m} \Delta\, Xj \times Wj^{\hat{a}} \times \exp(-\hat{b} \times cij)\right)^{-1} \qquad (27)$$

where Δ Xj is a measure of space in j.

In the case of a regular system of zones the Δ Xj terms disappear.

However, even this is relatively crude and ignores the more serious effect of aggregational variability whereby there are also an infinate number of alternative aggregations of points to the microstate zones used. Associated with these aggregations are changes in all the system constraints since they are only relative to a particular aggregation. Certainly the physicist never had to face this problem. It is apparent then that the entropy-maximising approach cannot incorporate very much by way of geographical information and it represents only a first stage of the model building process.

It is suggested that in order to become geographically relevant spatial model building has to have three distinct stages. The first stage concerns the specification of model form, and certainly entropy-maximising model building provides a good first attempt at model form. The second stage concerns the introduction of scale and aggregation effects, this is very important since these effects are implicit in all spatial data and they need to be explicitly controlled. The third stage concerns the introduction of macro spatial structure into the models, for example by replacing global deterrence functions by sets of functions defined over limited spatial domains. These stages are not independent of each other and modifications have to be made to the stage one models to reflect stages two and three. The alternative non-geographical approach is to regard model building as being a stage one all or nothing statistical process. However, there does seem to be a more geographical solution if only we would look for one.

(iii) Implications for the Planner

Professional planners are mainly interested in application studies. Does a particular technique work? What can be done to improve it? Often they can afford little time for experimentation and research in a field which is dominated by academic, theoretical, and mathematical approaches but with very little in the way of empirical study and evaluation. It is an area of overlap between geography and planning but very little traffic in ideas seems to be taking place. In fact the planner is often in the most unfortunate position of having to develop operational versions of models which were developed for theoretical purposes and which have not been properly evaluated.

At present the descriptive power of these shopping models is inadequate and their use in impact and forecasting studies needs to be carefully qualified. They are essentially first generation, elementary models. The provision of trip and cash flow data is urgently needed so that descriptive models can be built as well as to provide a framework for the update surveys and the monitoring that should accompany all impact and forecasting exercises. Only then can a real empirical assessment be made of how well these models perform and attempts made to develop new models based on better principles. In particular, the introduction of location-allocation versions offers considerable opportunity for the planning of retail provision. It is also necessary, whenever possible, to include confidence intervals on parameter values and on model forecasts even if they are embarrassingly large since these are the limits of our current level of understanding about the real world. These confidence intervals are also a measure of the degrees of control a planner may have over a future situation.

BIBLIOGRAPHY

Batty, M. (1971) Exploratory calibration of retail location models using search by golden section, *Environment and Planning,* 3,411-432.

Batty, M. (1974) Spatial entropy, *Geographical Analysis,* 1-31.

Batty, M. and S. Mackie (1972) The calibration of gravity, entropy and related models of spatial interaction, *Environment and Planning,* 4, 205-233.

Batty, M. and A. Saether (1972) A note on the design of shopping models, *RTPI Journal,* 58, 303-306.

Baxter, R. (1972) Entropy techniques for spatial interaction models, *Land use and Built Form Studies, Cambridge. Technical Notes Series B,* No.3.

Baxter, R. (1973) Entropy maximising models of spatial interaction, *Computer Applications,* 1, 57-71.

Berkshire County Council Self-Completion Shopping Diary, 1974.

Berry, B.J.L. (1965) The retail component of the urban model, *Journal of the American Institute of Planners,* 31, p.150.

Berry, B.J.L. (1967) *The geography of market centres and retail distribution,* (Prentice-Hall).

Bruce, A. and L. Dawes (1971) Shopping in Watford, *Report by the Building Research Station,* Garston.

Census of Distribution 1951, 1961, 1966 (incomplete), 1971; (HMSO: London).

Chilton, R. and R.R.W. Poet (1973) An entropy maximising approach to the recovery of detailed migration patterns from aggregate census data, *Environment and Planning,* 5, 135-146.

Cliff, A.D. and J.K. Ord (1973) *Spatial autocorrelation,* (Pion: London).

Converse, P.D. H.W. Huegy and R.W. Mitchell, *Elements of Marketing,* (Prentice Hall, 1930; 7th Edition 1965).

County Surveyor Society Trip Rate Data Bank, a copy is deposited with the SSRC Survey Library, University of East Anglia.

Davies, R.L. (1970) Variable relationships in central place and retail potential models, *Regional Studies,* 4, 49-61.

Davies, R.L. (1974) *Patterns and profiles of consumer behaviour,* (Department of Geography Research Series No. 10, University of Newcastle).

Dixon, L.C.W. (1972) *Non-Linear Optimisation,* (EUP: London).

Durham County Council Household Survey, *Going Shopping*, (1974).

Evans, A.W. (1971) The calibration of trip distribution models with exponential or similar cost functions. *Transportation Research*, 5, 15-38.

Family Expenditure Survey, published annually since 1953, (HMSO: London).

Gibson, M. and M. Pullen (1972) Retail turnover in the East Midlands: a regional application of a gravity model, *Regional Studies*, 6, 183-196.

Goldfeld, S.M. and R.E. Quandt (1965) Some tests for homoscedasticity, *Journal American Statistical Association*, 60, 539-554.

Gould, P. (1972) Pedagogic review, *Annals of the Association of American Geographers*, 62, 689-700.

Harris, B. (1968) Quantitative models of urban development, in: *Issues in Urban Economics*, edited H.S. Perloff and L. Wingo, (Resources for the Future: Johns Hopkins Press).

Haydock Report (1964) *Regional Shopping Centres*, A planning Report on N.W. England, Part One 1964; Part Two, *A Retail Market Potential Model*, (Manchester University, Department of Town and Country Planning).

Hill, D.M. (1965) A growth allocation model for the Boston Region, *Journal of the American Institute of Planners*, 31, 111-120.

Huff, D.L. (1962) A note on the limitations of intra-urban gravity models, *Land Economics*, 38, 64-66; and 39, 81-89.

Huff, D.L. (1964) Defining and estimating a trading area, *Journal of Marketing*, 28, 37-48.

Hyman, G.M. (1969) The calibration of trip distribution models with exponential or similar cost functions, *Transportation Research*, 5, 15-38.

Kilsby, D.J.E., J.S. Tulip and M.R. Bristow (1973) The attractiveness of shopping centres, *CURR Occasional Paper*, No.12.

Lakshmanan, T.R. and W.G. Hansen (1965) A retail market potential model, *Journal of the American Institute of Planners*, 31, 134-143.

Lewis, J.P. and M.J. Bridges (1974) The two-stage household shopping model used in the Cambridge Sub-region study, *Regional Studies*, 8, 287-297.

Lowry, I.M. (1964) *A model of metropolis*, (Rand Corporation, RW - 4125 - RC).

Mackett, R.L. (1973) Shopping in the city: the application of an intra-urban shopping model to Leeds, *Department of Geography, Leeds University, Working Paper*, 30.

Mountcastle, G.D., J.C.H. Stillwell and P.H. Rees (1974) A users guide to a program for calibrating and testing spatial interaction models, *Department of Geography, Leeds University, Working Paper*, 79.

Murray, W. and M.B. Kennedy (1971) Notts/Derbys: A shopping primer, *RTPI Journal*, 58, 211-215.

Nedo (1970) *Urban models in Shopping studies*, (Distributive Trades EDC, National Economic Development Office, London).

Openshaw, S. (1973) Insoluable problems in shopping model calibration when the trip pattern is not known, *Regional Studies*, 7, 367-371.

Openshaw, S. (1974) Calibration and behaviour of some shopping models, *Proceedings of PTRC Summer Conference, University of Warwick*.

Openshaw, S. (1977) SIMS - user manual, *Department of Town and Country Planning, Newcastle University, Working Paper*, 3.

Openshaw, S.and C.J. Connolly (1975) *A family of shopping location-allocation models*, (mimeo).

Openshaw, S. and C.J. Connolly (1976) *An empirical study of some families of continuous and discontinuous deterrence functions for maximum performance spatial interaction models*. (Paper presented at 9th Annual RSA Conference, London, Sept. 1976).

e, A.S. (1969) Gravity models in Town Planning - uses in retailing, *Gravity Modes in Town Planning*, (Lanchester Polytechnic).

eilly, W.J. (1931) The Law of retail gravitation, (G.P. Putman: New York).

Rodgers, D.S. (1974) Bretton, Peterborough: the impact of a large edge-of-town supermarket, *Retail Outlets Research Unit, Manchester Business School. Research Report*, 9.

ilvey, S.D. (1970) *Statistical Inference*, (Penguin: London).

mith, A.P. (1973) Retail allocation models: an investigation into problems of application to the Rotherham sub-region, *Department of Geography, Leeds University, Working Paper*, 50.

Stouffer, S.A. (1940) Intervening opportunities: a theory relating mobility and distance, *American Sociological Review*, 5, 845-867.

Styles, B.J. (1969) Principles and historical development of the gravity concept, in: *Gravity Models in Town Planning*, (Lanchester Polytechnic).

Taylor, P.J. (1975) Distance decay models in spatial interactions, *CATMOG*, 2 (Geo Abstracts Ltd: University of East Anglia).

Wade, B.F. (1973) *Greater Peterborough Shopping Study: Technical Report*, (PRAG: London).

Williams, I. (1974) A comment on the derivation of calibration criteria by Batty and Mackie, *Environment and Planning*, 6, 603-605.

Wilson, A.G. (1967) A statistical theory of spatial distribution models, *Transportation Research*, 1, 253-269.

Wilson, A.G. (1970) *Entropy in urban and regional modelling*, (Pion: London).

Wilson, A.G. (1971) A family of spatial interaction models and associated developments, *Environment and Planning*, 3, 1-32.

Wilson, A.G. (1974) *Urban and regional models in geography and pl ning*, (Wiley: London).

Wilson, A.G. (1974) Retailers profit and consumer welfare in spatial inter action shopping models, *Department of Geography, Leeds Uni versity, Working Paper*, 78.